Science Museum

Submarine Telegraphy
The Grand Victorian Technology

Bernard S Finn

Curator, National Museum of History & Technology
Smithsonian Institute

1973

Printed for Her Majesty's Stationery Office by
Eyre & Spottiswoode Ltd
Thanet Press, Margate

Price 33p

Dd 228963 K32 6/73

Since the discovery of Columbus nothing has been done in any degree comparable to the vast enlargement which has thus been given to the sphere of human activity.

The Times 6 August 1858

. . . the great work of the age.

New York Herald 20 April 1857

Foreword

The author of this booklet, Dr Bernard S. Finn, is a curator at the National Museum of History and Technology of the Smithsonian Institution in Washington. While based here for a sabbatical year researching into the history of submarine telegraphy, he planned a Science Museum special exhibition and wrote this booklet to accompany it.

Dr Finn is no stranger to us, and we were very pleased when he chose to spend his year here at the Science Museum. We have enjoyed the opportunity for co-operation which his visit has provided and are grateful to him for his efforts on our behalf. I would like to think that more scholars of technological history will come in the future and base themselves here in the same sort of way.

Margaret Weston
Director

1 Coiling the cable on board the *Agamemnon*, 1857

THE COILING OF THE ATLANTIC TELEGRAPH CABLE ON BOARD H.M.S. "AGAMEMNON."

Preface

In 1950 the Science Museum sponsored a special exhibition marking the centenary of submarine telegraphy. At that time radical changes were taking place in the industry: a new synthetic insulating material, polythene, had replaced the traditional gutta percha, and underwater amplifiers were being used on some of the shorter cables. It was clear that the whole system would eventually be affected, but it was still too early to tell just how drastic the consequences would be.

Today these and other innovations have produced revolutionary changes in the industry. As a result, most of the old terminal stations have been abandoned, and the apparatus has been dispersed or scrapped.* As I write this we are planning to open in the Science Museum in July 1973 another special exhibition in which an attempt will be made to take a new look at the old technology. I think we may be able to see its importance more clearly now that it has gone. And I believe that this may help us to understand better the impact of the new technology that has taken its place. The themes of this booklet are the same as those of the exhibition. Hopefully it will prove interesting to those who have not seen the exhibition as well as to those who have.

It has been a rewarding experience for me to receive the assistance of a large number of people on this project. In the Science Museum these include: Margaret Weston, Don Chilton, Lesley West, Brian Bowers, Keith Geddes, Brian Owen, Peter Scott, Michael Preston, Christine Heap, Anita McConnell, and the staffs of the Photographic Studio and the Library. Elsewhere special thanks go to: Jean Goodship at the Institution of Electrical Engineers; Ken Baglehole, Skip Slape, Nicky Parker and Malcolm Lumb at Cable and Wireless; Phil Sykes and Harry Stocker at Standard Telephones & Cables; Major J. K. Zieleznik at the Post Office Museum; and Anastasia Atsikoudas and Elliot Sivowitch at the Smithsonian. Finally invaluable commentary and advice have come from Gerald Garratt, Ken Haigh and Field Curry.

Bernard S. Finn London, March 1973

*Two stations have been preserved as museums, at Hearts Content, Newfoundland, and at Orleans, Massachusetts. Instruments have been preserved in them and in the collections of some technical museums, notably at the Science Museum in London, and at the Smithsonian's National Museum of History and Technology in Washington.

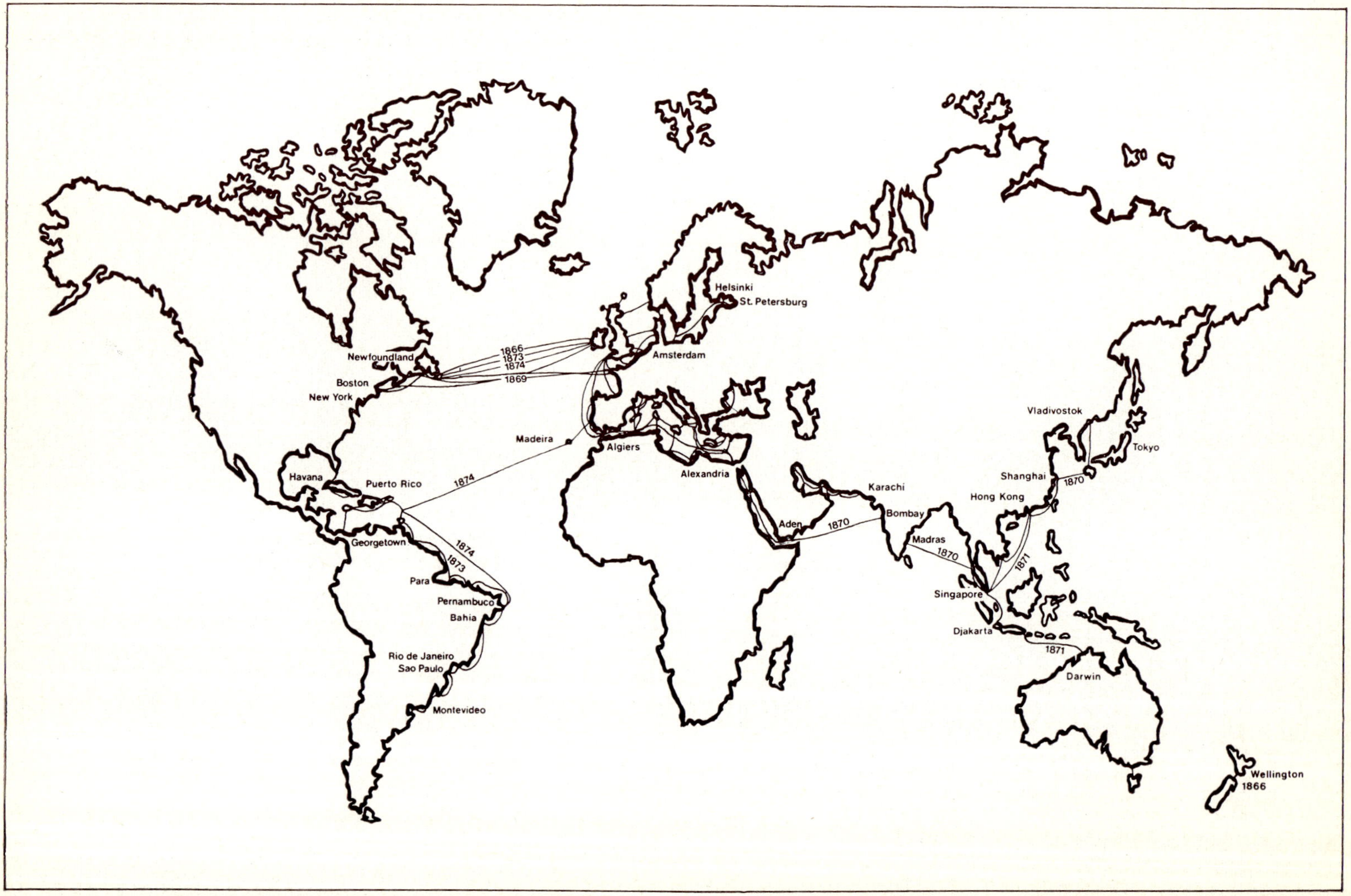

2 Major cables in operation in 1875

3

Introduction

The confidence that Victorian Britain felt in its technology, in its commerce and in its political system could hardly have been better expressed than through the development of submarine cables. The technical skills and the capital were obtained in Britain, and Britain was the centre from which lines reached out to embrace the globe. Distances from London that had formerly been measured in weeks or months were encompassed in minutes, with profound consequences for politics and commerce. Some of the effects had been anticipated by the original promoters; others could hardly have been foreseen in a world that was undergoing such vast and general social change. Even now it is difficult to measure the extent of the alterations that have come about because of these slender but durable wires.

A special technology was developed with the cables, including unique methods of signalling and detecting that were necessary because of the long distances covered and the electrical characteristics of the lines. This technology (or at least a large part of it) became obsolete in the 1950s and 1960s when amplifiers were developed which could be submerged with the cables and when communications satellites began to appear. Old cables, old instruments, old stations were abandoned. In their place came a much improved technology with new opportunities and new uncertainties.

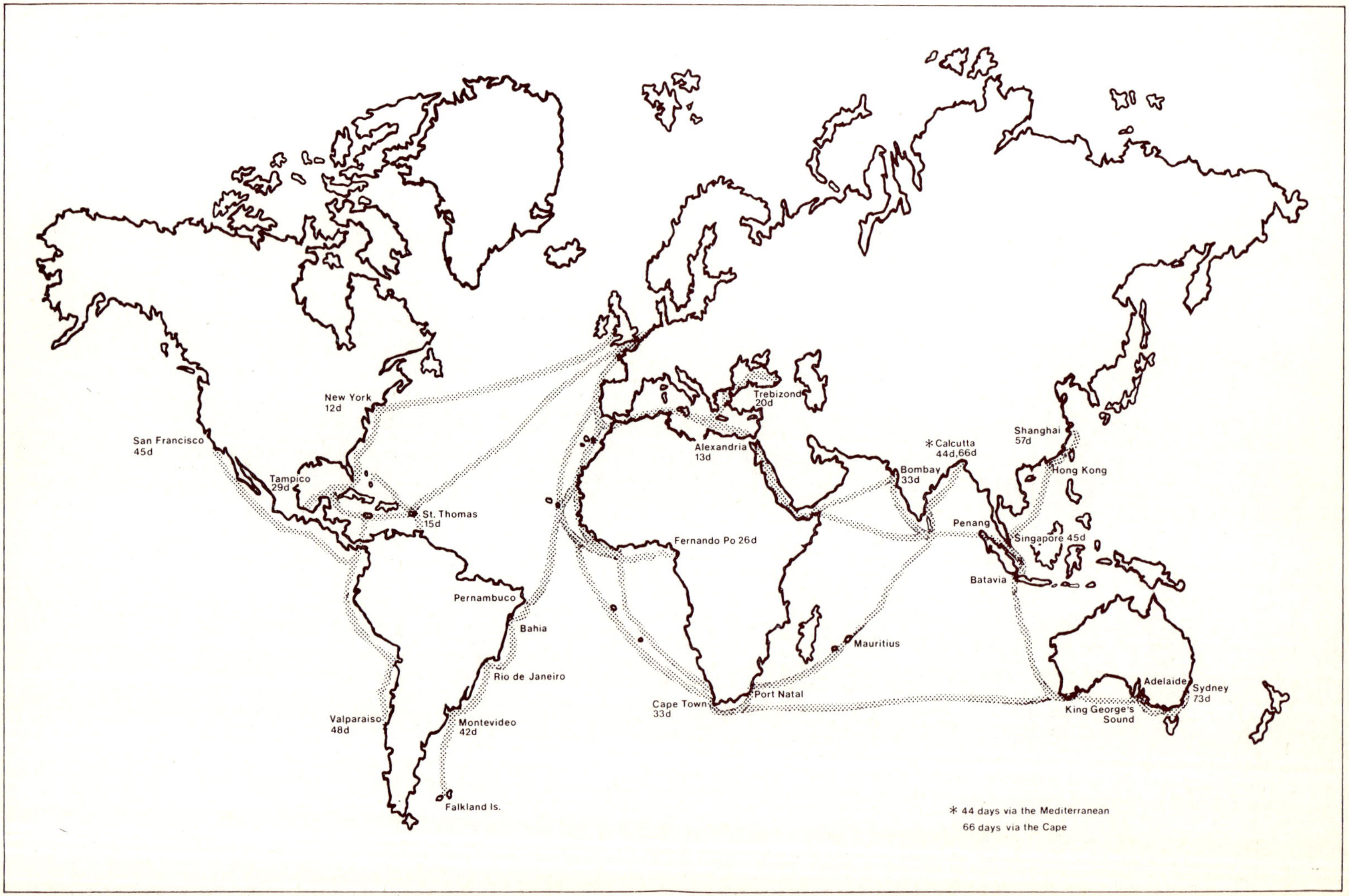

4 Overseas postal routes in 1852 showing in days the average time for mail to reach England

5 Cooke and Wheatstone receiver

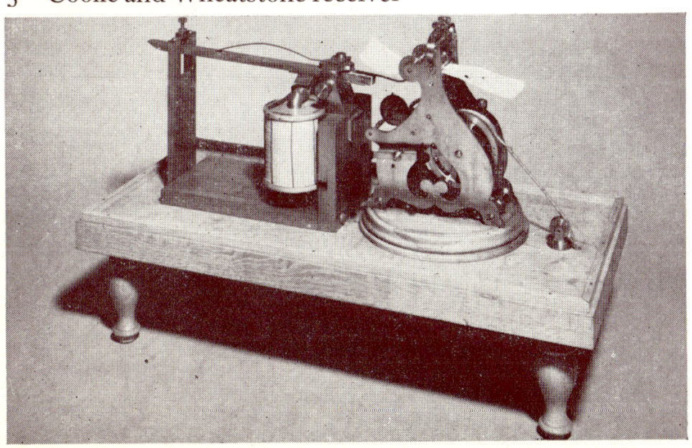

6 Morse receiver

1 Preconditions

The rapid growth of submarine telegraphy was possible because the necessary incentives, capital and technology were available. By the mid-nineteenth century all were present in Britain in sufficient measure so that success could be achieved if a good promotional effort was made.

The primary incentives were economic. At mid-century British trade was expanding rapidly as the country took advantage of its position as the first nation to industrialize. By 1870 exports would exceed 20 per cent of national income, and British investments abroad would reach £1000 million. This type of activity demanded quick and reliable communications. The London banker needed to know the state of the stock exchanges in Paris and New York. The Liverpool shipper gained considerably if he could communicate with his ship's captain in Singapore about the commodities available and their prices and thus make judgements at the time of loading instead of two or three months before.

As other countries developed industrially the Empire became more and more important to British trade. India, for instance, was absorbing almost half of Britain's cotton exports in the 1870s. In this situation it was politically essential that contact between the colonies and the mother country should be as close as possible. Realization of this need was expressed specifically by many who invested in cable companies. They felt that they were investing in the future of their country, and it is not surprising to find the Government itself giving support to several cable enterprises.

Capital was available in a prosperous British Economy, and hundreds of thousands of pounds would easily be raised if the risk seemed reasonable. Significantly, this was not generally true elsewhere: in spite of strong reasons for doing so, American investors contributed very little to the early Atlantic cables.

No matter what the motivations, technically it would have been impossible to lay cables successfully in, say, 1840. But a decade later a number of essential developments had taken place, making such ventures reasonable. Three technical achievements were of special

8 Advertisement by the Gutta Percha Company, c.1849

importance: long distance electric telegraphy, the use of steam engines in large ships, and the discovery of a proper insulating material.

Although still in its early stages, telegraphy on land had become well established by 1850. The Cooke and Wheatstone single-needle receiver and the Morse embossing instrument could both operate over moderately long distances and had been demonstrated through lines more than a thousand miles long. It was not unreasonable to suppose that the width of the Atlantic could be handled with minor adjustments.

Steam ships were essential for laying cable at a controlled rate in a fairly straight line. For the early, short cables a number of possible vessels were available. However, for the first Atlantic cable attempt in 1857 a much larger carrying capacity was required, in fact too much for any single ship then afloat. Two were used: the 3200 ton *Agamemnon*, built for the Royal Navy in 1852, and the 4600 ton *Niagara* built for the US Navy in 1855. With some internal structural alterations to provide room for cable tanks, each was capable of carrying over a thousand miles of cable. Converted vessels continued to be used during the rapid expansion of submarine telegraphy in the years that followed, and it was not until 1874 that the first purpose-built cable ship was constructed.

Good insulation was needed. Fortunately gutta percha, a natural plastic material from the sap of certain trees in and around Malaya had been introduced to Britain in 1843. In the next few years techniques for working it were developed, including the ability to extrude it around a copper wire.

7 *Agamemnon* and *Niagara* in mid-ocean, 1858

9 John W. Brett

10 Jacob Brett

11 Charles T. Bright

12 Cyrus Field

9 John W. Brett (1805-1863) had been successful as a 'dealer in curiosities.' He supported his brother on the early channel cables, then went on to promote lines in the Mediterranean, losing much money in an unsuccessful attempt from Sardinia to Africa in 1855. He invested heavily in the first Atlantic cable.

10 Jacob Brett (1808-1897) promoted the idea of a cross-channel telegraph as early as 1845, hoping to use his modification of R. E. House's printing instrument on it. He failed to capitalize on his early pioneering efforts and died in comparative poverty.

11 Charles T. Bright (1832-1888) supervised laying of the first successful cable to Ireland in 1853, the Atlantic cables of 1857 and 1858, the Persian Gulf cable of 1864, and a series of cables in the West Indies starting in 1869 (a venture that was financially disastrous). He was important in the development of many basic techniques.

12 Cyrus Field (1819-1892) devoted a considerable portion of his time and fortune to the Atlantic cable. He was unsuccessful in persuading the U.S. government to lay cables along the East Coast during the Civil War, and he was an early advocate of an American Pacific cable.

13 Willoughby Smith (1828-1891) was an electrician with the Gutta Percha Company and with TC&M, and also later with the instrument firm of Elliott Brothers. He made important contributions to insulation techniques and testing procedures.

14 Frederic N. Gisborne (1824-1892) was a telegraph engineer who promoted the Newfoundland Electric Company and laid the first undersea cable in North America, between Prince Edward Island and New Brunswick, in 1852. Deep in debt and unable to continue, he sold his concession to the group headed by Cyrus Field.

15 William Thomson (later Lord Kelvin) (1824-1907), Professor of Natural Philosophy at the University of Glasgow, was an adviser to the Atlantic Cable project, accompanying all four expeditions. He developed the theory of signal transmission, designed several instruments for sending, receiving and testing, and devised more efficient methods of deep-sea sounding.

16 John Pender (1816-1896), a Manchester merchant who invested in the first Atlantic cable, organised TC&M and headed several companies to promote cables reaching westward to South America and eastward to the Pacific.

13 Willoughby Smith

14 Frederic N. Gisborne

15 William Thomson

16 John Pender

17 By 1857 a number of cables joined England to the mainland, and extensions south and eastward had been made through the Mediterranean

18 Drawing of 1850 Channel cable with lead weight, and original weight from Science Museum collections

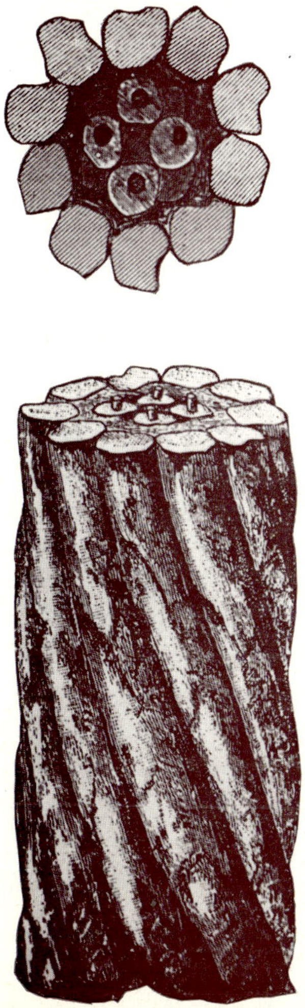

19 1851 Channel cable, picked up in 1859

2 The Pioneers

Although they were confident of success, brothers John W. and Jacob Brett had considerable difficulty in persuading friends to take advantage of a concession they had obtained from the French Government in 1849 for a cross-channel telegraph line. They finally managed to raise £2000 and contracted with the Gutta Percha Company for a single strand of copper wire covered with gutta percha. Lead weights were to be used to hold the cable down, and lead tubes were to protect it near the shores. The cable was laid during the day of 28 August 1850, and signals were exchanged that evening for a few hours before a fisherman's anchor broke the connection.

In 1851 assistance was obtained from railway engineer T. R. Crampton, who designed the cable and supervised its construction and laying. Gutta percha was again used for the insulation, but this time conductors for four separate circuits were included and the whole was protected by an outer wrapping of iron armour wires. The armouring contract was given to wire rope manufacturers Wilkins and Weatherly, but it was taken over by R. S. Newall & Company on the grounds that the manufacturing process infringed Newall's 1840 patent for a wire rope with a soft core. The cable was laid on 25 September 1851 and remained in use for many years.

The idea of spanning the Atlantic had been proposed even before connections were made across the English Channel. The greatest technical problem envisaged was the mechanical one of laying the cable. The strain on the machinery produced by a cable hanging in water two miles deep was going to be considerable; furthermore, if the cable broke grappling techniques could not be relied upon to retrieve the lost end. The promotional problem was to persuade enough people to risk money in the face of these difficulties.

The chief persuader was Cyrus Field, an American who by the age of 34 had made a fortune in the wholesale paper business and was looking for a new project to excite his interest. A chance encounter introduced him to Frederic N. Gisborne, who was deeply in debt as the result of an attempt to run a telegraph line across Newfoundland and a cable

20 Armouring the Atlantic cable, 1857

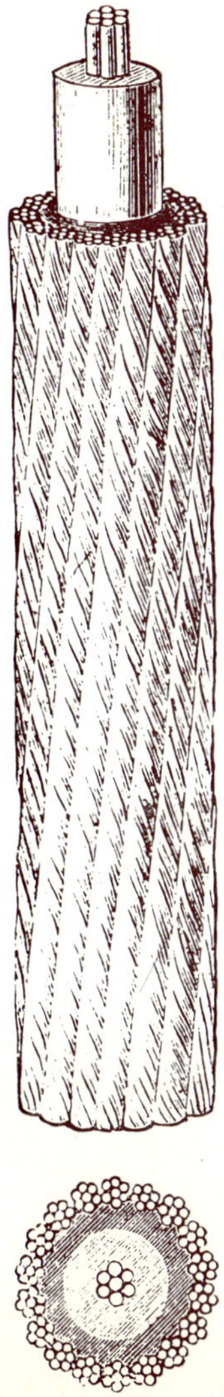

21 Atlantic cable of 1857

link to the North American mainland. Field and four associates proceeded to spend about $1,000,000 (£200,000) of their own money completing this line. At the same time they obtained landing concessions for a transatlantic cable, and in 1855 Field made the first of many trips to England to promote the venture. Most of £350,000 needed for the Atlantic Telegraphy Company was raised, as expected, from the bankers and merchants of Manchester, Liverpool and London in the form of £1000 shares. Subsidies payable only in the event of success were guaranteed by the United States and British governments. The core of the 2500 nautical miles of cable was manufactured by the Gutta Percha Company; half was armoured by Glass, Elliott & Company at Greenwich and the other half by Newall, all in the incredibly short time of four months. As events proved, a little more time spent on the design of the cable and on quality control would have been well worth while.

Starting on 6 August 1857 the frigate *Niagara* began laying cable from Valentia, Ireland. It broke after three-quarters of an hour, so they started again. This time all went smoothly until 334 nautical miles of cable had been deposited in the ooze of the ocean floor, now two miles beneath the ship. Then a brake was incorrectly applied to the paying-out mechanism and the line parted, abruptly ending the expedition. The ships returned to port and were unloaded. An additional 700 nautical miles of cable were ordered from the Gutta Percha and Glass, Elliott companies, in preparation for another try the next year.

On 10 June 1858 the *Agamemnon* and the *Niagara* set out again, burdened with some 3000 miles of cable. This time they were to meet at mid-Atlantic, splice the ends, and proceed in opposite directions. After enduring for a week one of the worst storms ever recorded on the North Atlantic, in which the *Agamemnon* nearly foundered, they arrived at the appointed place, and the tedious paying-out began on June 26. Electrical test-sets on both ships kept constant vigil over the condition of the cable. After about six miles of cable had disappeared

22 Loading cable on the *Agamemnon*, 1857

23 Tangled cable on the *Agamemnon* during the storm

into the water there was a break. The ships returned to the starting point and began anew. This time 80 miles were laid before the line parted. Another attempt produced another failure, losing about 200 miles of cable. The ships then returned to port, under the assumption that too much cable had been lost to proceed. In fact there was a little more than enough left, and the Company directors decided it was worth making one last effort. The ships returned to mid-ocean and spliced ends on July 29. On August 5 the *Agamemnon* steamed into Valentia while the *Niagara* was arriving at the Bay of Bull's Arm in Trinity Bay, Newfoundland.

After the cable had been landed there was a pause. A series of tests were made to determine the best operating conditions. These indicated that there were some major problems which had not been completely anticipated. Finally – although the situation was far from satisfactory – preparations were made for a formal inauguration of the line, and on August 13 the Queen and the President exchanged messages.

The apparent success of the cable touched off a series of celebrations in America. Coastal cities sponsored parades, dinners and speeches. Many inland cities, with no immediate stake in the commercial aspects of the new communications link, set aside time for prayer and fireworks. Souvenirs from the expedition found a quick and eager market. The firm of Tiffany's, in New York, purchased the remainder of the cable from the *Niagara*, cut part of it up in four-inch pieces, and had no difficulty in selling them at fifty cents each when accompanied by a certificate of authenticity signed by Cyrus Field. Other sections were made into umbrella handles, canes and watch fobs. In England reaction was a bit more restrained, though the newspapers widely proclaimed this the greatest of man's scientific and technical achievements. Charles Bright, then 26, was knighted for his role as the engineer who supervised the laying operation, and others received lesser recognition.

Unfortunately, the formal inauguration was followed by a period of even greater uncertainty. The condition of the cable deteriorated

224 L'ILLUSTRATION, JOURNAL UNIVERSEL.

MANIFESTATION POPULAIRE A NEW-YORK, A L'OCCASION DE L'ACHÈVEMENT DU TÉLÉGRAPHE TRANSATLANTIQUE.

24 New York celebration in 1858 after messages between Queen and President had been exchanged

25 One of Whitehouse's induction coils

rapidly, and signals were distinguished only with difficulty. A few messages were sent in August, including cancellation of an order to despatch two regiments from Canada (a consequence of the amelioration of the situation in India). This alteration in plans is said to have saved the British Government some £50,000. In September the receiving apparatus flickered in mostly incomprehensible fashion. After October 20 it responded not at all. The cable was never opened to commercial traffic.

The post mortem examination was as detailed as practical, considering the general inaccessibility of the corpse. The conclusions reached were tentative, but are as good as any we can make today. The cable had been hastily manufactured, tossed about at sea, and stored dry on a dock through the winter (later practice would protect the gutta percha from heat and air by storing the cable under water, both at the factory and on ship). Any weak spots that had formed then broke down under the 3000-volt pulses from the induction coils which the company's electrician, Wildeman Whitehouse, inadvisedly used in attempts to activate ordinary receiving instruments.

Electrical characteristics of submarine telegraph cables

Long submarine cables differed electrically from ordinary land telegraph lines in two principal ways. First, no relays (amplifiers) could be used along the length; therefore the received signal was extremely feeble. Second, because the cable was surrounded by a conductor (sea water) it acted like a long capacitor storing up the electrical pulse and then gradually (relatively speaking) releasing it; a sharp signal at one end appeared smeared out at the other, thus severely restricting the number of pulses that could be transmitted and identified in a given time.

To solve the first problem William Thomson (later Lord Kelvin) designed a sensitive mirror galvanometer. A very light-weight magnet and mirror assembly hung suspended at the centre of a coil. The

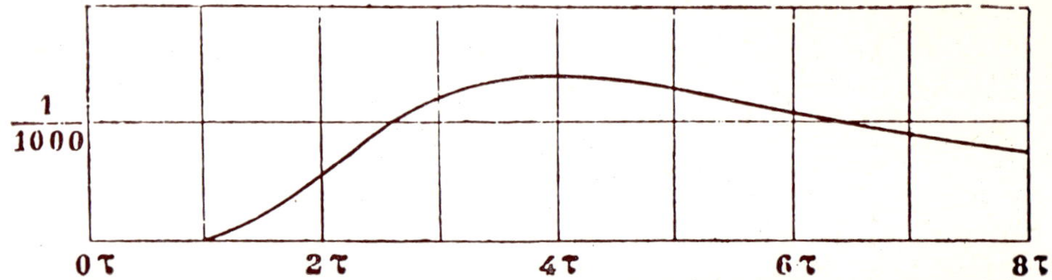

26 Signal pulse as it would appear at the far end of cable *from a contemporary publication*

27 Arrangement of Thomson's mirror galvanometer

received signal was fed through the coil, creating a small magnetic field which turned the magnet and mirror on the suspending filament. A beam of light, reflected from the mirror onto a scale, would show even the slightest turn. The mirror galvanometer was first used on the 1858 cable when Whitehouse's system proved ineffective. Although later replaced as a received by other devices, it has been a standard instrument for testing purposes up to the present day.

In 1858 the solution to the second problem was simply to transmit very slowly. Later, various techniques would be introduced to compensate for the capacitive effect, with the result that substantial speeds could be achieved.

The indefatigable Field, and others, argued that far from being a failure the experience of 1858 clearly demonstrated not only that it was possible to lay a cable across the Atlantic but also that messages could indeed be transmitted through it. All that was needed was a little more attention to detail in the manufacture and a little more care in operation. Unfortunately, two years later £800,000 was lost on the much-heralded submarine cable from Suez to India, and it is not difficult to understand that investors may have felt somewhat dubious about the possibilities of long lines. Shorter cables, however, continued to be laid. Then in 1861 a Government report appeared which analysed the technology of cable manufacture and use and concluded in effect that the major problems had been solved.

In 1863 the Government, anxious to establish communications between England and India, decided to sponsor a cable up the Persian Gulf from Karachi to Fayal to tie in with land lines through Persia and Turkey to Europe. After making careful surveys, the Director General of the Indian Telegraph Department, Patrick Stewart, went to England and engaged Charles Bright as Consulting Engineer. Full advantage was taken of past experience, and the cable was successfully laid in 1864. Unfortunately, there were continual difficulties

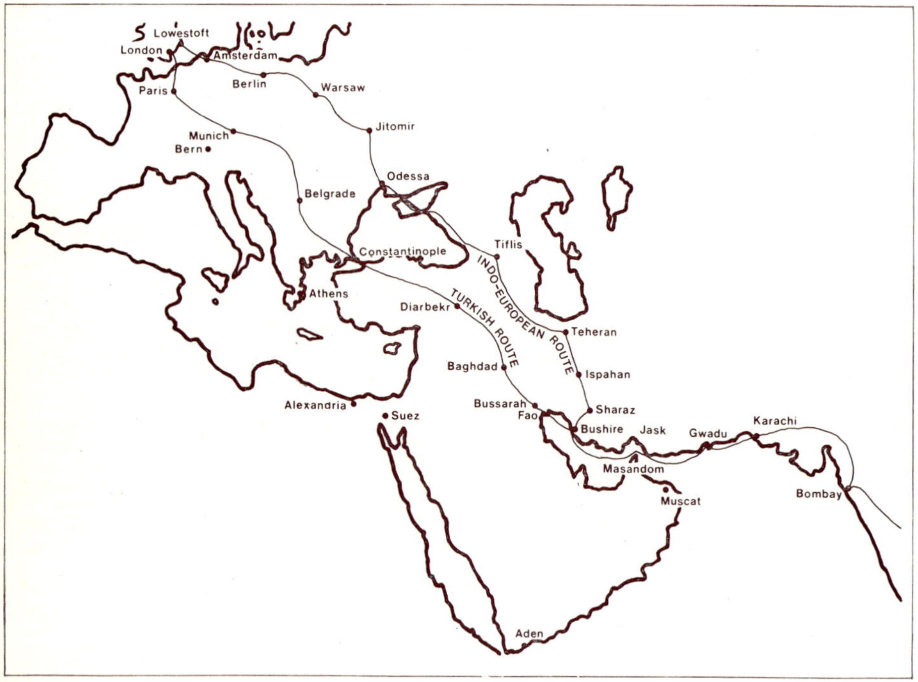

28 Persian Gulf cable to India, 1864

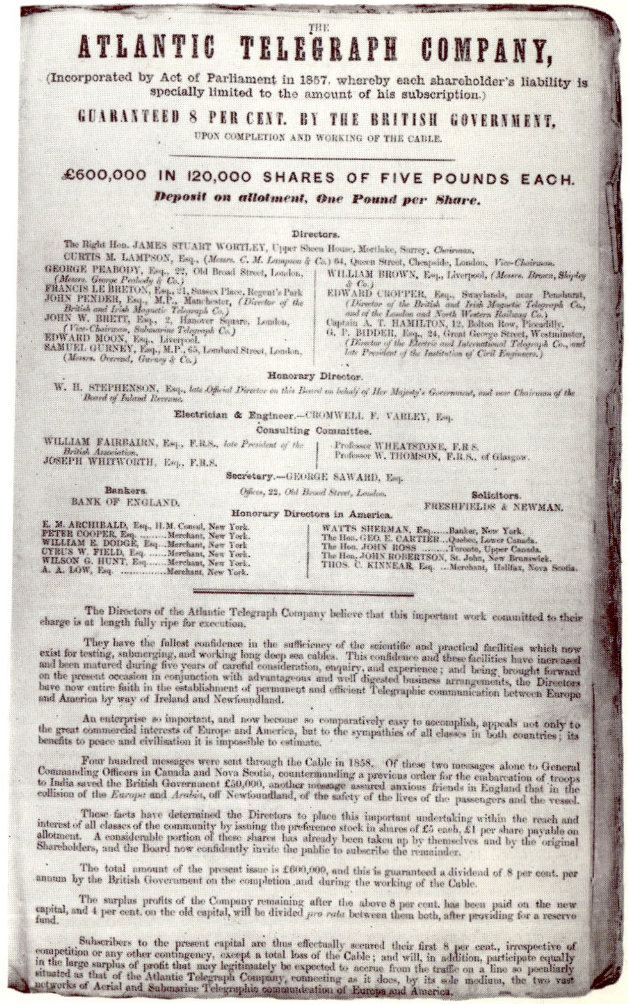

THE

ATLANTIC TELEGRAPH COMPANY,

(Incorporated by Act of Parliament in 1857, whereby each shareholder's liability is specially limited to the amount of his subscription.)

GUARANTEED 8 PER CENT. BY THE BRITISH GOVERNMENT,

UPON COMPLETION AND WORKING OF THE CABLE.

£600,000 IN 120,000 SHARES OF FIVE POUNDS EACH.

Deposit on allotment, One Pound per Share.

Directors.

The Right Hon. JAMES STUART WORTLEY, Upper Sheen House, Mortlake, Surrey, *Chairman.*

CURTIS M. LAMPSON, Esq., (*Messrs. C. M. Lampson & Co.*) 64, Queen Street, Cheapside, London, *Vice-Chairman.*

GEORGE PEABODY, Esq., 22, Old Broad Street, London, (*Messrs. George Peabody & Co.*)

FRANCIS LE BRETON, Esq., 21, Sussex Place, Regent's Park

JOHN PENDER, Esq., M.P., Manchester, (*Director of the British and Irish Magnetic Telegraph Co.*)

JOHN W. BRETT, Esq., 2, Hanover Square, London, (*Vice-Chairman, Submarine Telegraph Co.*)

EDWARD MOON, Esq., Liverpool.

SAMUEL GURNEY, Esq., M.P., 65, Lombard Street, London, (*Messrs. Overend, Gurney & Co.*)

WILLIAM BROWN, Esq., Liverpool, (*Messrs. Brown, Shipley & Co.*)

EDWARD CROPPER, Esq., Swaylands, near Penshurst, (*Director of the British and Irish Magnetic Telegraph Co., and of the London and North Western Railway Co.*)

Captain A. T. HAMILTON, 12, Bolton Row, Piccadilly.

G. P. BIDDER, Esq., 24, Great George Street, Westminster, (*Director of the Electric and International Telegraph Co., and late President of the Institution of Civil Engineers.*)

Honorary Director.

W. H. STEPHENSON, Esq., *late Official Director on this Board on behalf of Her Majesty's Government, and now Chairman of the Board of Inland Revenue.*

Electrician & Engineer.—CROMWELL F. VARLEY, Esq.

Consulting Committee.

WILLIAM FAIRBAIRN, Esq., F.R.S., *late President of the British Association.*

JOSEPH WHITWORTH, Esq., F.R.S.

Professor WHEATSTONE, F.R.S.

Professor W. THOMSON, F.R.S., of Glasgow.

Secretary.—GEORGE SAWARD, Esq.

Offices, 22, Old Broad Street, London.

Bankers.
BANK OF ENGLAND.

Solicitors.
FRESHFIELDS & NEWMAN.

Honorary Directors in America.

E. M. ARCHIBALD, Esq., H.M. Consul, New York.
PETER COOPER, Esq. ....Merchant, New York.
WILLIAM E. DODGE, Esq...Merchant, New York
CYRUS W. FIELD, Esq. ......Merchant, New York.
WILSON G. HUNT, Esq.......Merchant, New York.
A. A. LOW, Esq. ....Merchant, New York.
WATTS SHERMAN, Esq......Banker, New York.
The Hon. GEO. E. CARTIER...Quebec, Lower Canada.
The Hon. JOHN ROSS ....Toronto, Upper Canada.
The Hon. JOHN ROBERTSON, St. John, New Brunswick.
THOS. C. KINNEAR, Esq. ...Merchant, Halifax, Nova Scotia.

The Directors of the Atlantic Telegraph Company believe that this important work committed to their charge is at length fully ripe for execution.

They have the fullest confidence in the sufficiency of the scientific and practical facilities which now exist for testing, submerging, and working long deep sea cables. This confidence and these facilities have increased and been matured during five years of careful consideration, enquiry, and experience; and being brought forward on the present occasion in conjunction with advantageous and well digested business arrangements, the Directors have now entire faith in the establishment of permanent and efficient Telegraphic communication between Europe and America by way of Ireland and Newfoundland.

An enterprise so important, and now become so comparatively easy to accomplish, appeals not only to the great commercial interests of Europe and America, but to the sympathies of all classes in both countries; its benefits to peace and civilisation it is impossible to estimate.

Four hundred messages were sent through the Cable in 1858. Of these two messages alone to General Commanding Officers in Canada and Nova Scotia, countermanding a previous order for the embarcation of troops to India saved the British Government £50,000, another message assured anxious friends in England that in the collision of the *Europa* and *Arabia*, off Newfoundland, of the safety of the lives of the passengers and the vessel.

These facts have determined the Directors to place this important undertaking within the reach and interest of all classes of the community by issuing the preference stock in shares of £5 each, £1 per share payable on allotment. A considerable portion of these shares has already been taken up by themselves and by the original Shareholders, and the Board now confidently invite the public to subscribe the remainder.

The total amount of the present issue is £600,000, and this is guaranteed a dividend of 8 per cent. per annum by the British Government on the completion and during the working of the Cable.

The surplus profits of the Company remaining after the above 8 per cent. has been paid on the new capital, and 4 per cent. on the old capital, will be divided *pro rata* between them both, after providing for a reserve fund.

Subscribers to the present capital are thus effectually secured their first 8 per cent., irrespective of competition or any other contingency, except a total loss of the Cable; and will, in addition, participate equally in the large surplus of profit that may legitimately be expected to accrue from the traffic on a line so peculiarly situated as that of the Atlantic Telegraph Company, connecting as it does, by its sole medium, the two vast networks of Aerial and Submarine Telegraphic communication of Europe and America.

29 Prospectus issued in 1865

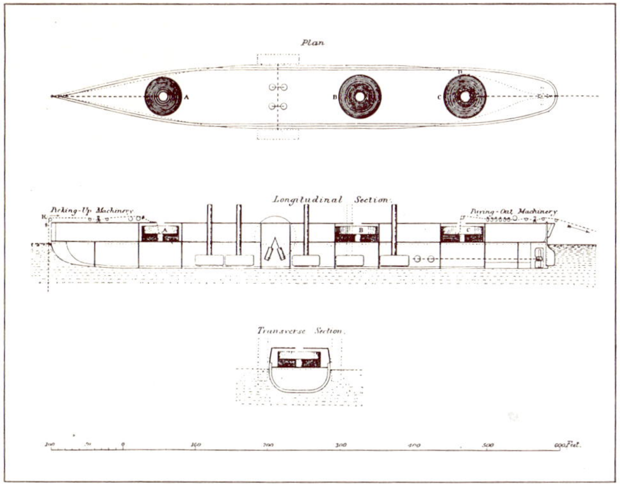

30 Arrangement of cable tanks and machinery on the Great Eastern, 1865-66

with the land lines, inefficient administration being aggravated by the occasional excesses of marauding tribes. Also, the governments of the countries which the lines crossed required the use of native clerks who were often unfamiliar with English, which compounded the problem of relaying messages. As a result, the cable did not receive either the business or the recognition that it deserved.

Meanwhile, the Atlantic Cable promoters also decided to try again. The American Civil War had reinforced the view that rapid channels of communication were essential across the Atlantic, and as the war drew to a close it became easier to act. £500,000 were sought, this time in shares of five pounds to attract more investors. John Pender, a Manchester merchant who had been a supporter of the 1858 venture, used his organizing talents to bring the Gutta Percha Company and Glass, Elliott together to form the Telegraph Construction and Maintenance Company (TC&M) in 1864. The new company made the main part of the cable, taking shares in the Atlantic Telegraph Company as part of their payment. Shore ends were manufactured by W. T. Henley. The *Great Eastern*, I. K. Brunel's magnificent white elephant that had failed economically as a passenger liner, was converted into a cable ship, and the full length was loaded into her hold. The ship left Valentia on 23 July 1865, and with only a few minor incidents paid out 1186 nautical miles of cable until in an attempt to correct a flaw there was a break. Attempts at recovery were unsuccessful, but came close enough to show that the cable could be retrieved with proper equipment. As a result, confidence was higher than ever. An additional £600,000 were raised under a new corporate name, the Anglo-American Telegraph Company, and on Friday, 13 July 1866, the *Great Eastern* set out again. This time there were no breaks, and the cable was landed at Hearts Content in Trinity Bay on the 27th. The ship returned to the point where the cable had been lost the previous year and eventually managed to retrieve it. A new length of cable was spliced on, and the line was completed to Hearts Content on September 8. There were then two Atlantic cables.

31 The *Great Eastern*

32 Recovering the 1865 cable

a 1850. English Channel. Solid copper wire, gutta percha.

b 1851. English Channel. Four single copper wires (each covered with two layers of gutta percha), tarred hemp, galvanized iron wires (for strength and protection)

c 1857-1858. Atlantic (deep sea section). Stranded copper wire, three layers of gutta percha, tarred hemp, stranded iron wires (for strength; ends of cable had heavier iron wires for protection).

d 1846. Persian Gulf. Segmented copper wire in copper tube, Chatterton's compound (tar, resin, gutta percha), four layers of gutta percha with Chatterton's compound between layers, tanned hemp (which did not act as an insulator—as tarred hemp did—and thus allowed defects in the gutta percha to show up earlier in tests), galvanized iron wires, two layers of hemp impregnated with Bright & Clark compound (pitch and tar to keep water from the iron wires, and silica to protect the core from boring insects).

e 1865-1866. Atlantic (deep sea section). Stranded copper wire, Chatterton's compound, three layers of gutta percha with Chatterton's compound between layers, tanned jute iron wires (galvanized in 1866 cable) each covered with tarred hemp (not tarred in 1866 to prevent sections from sticking together).

f 1926. Emden-Azores; loaded (intermediate deep sea section). Central copper wire with copper strips around it, permalloy tape, three layers of gutta percha with Chatterton's compound between layers, brass 'teredo' tape (to protect core from boring insects; not included in deep sea section), one layer tanned hemp, two layers of jute, tarred iron wires, two layers of jute.

g 1956. TAT-1; first trans-Atlantic telephone cable (deep sea section). Solid copper wire three copper tapes around it, polythene, outer conductor made of six copper tapes, copper teredo tape, taron tape, jute, steel wire, two layers of tarred jute. (In 1961 a new light-weight cable was introduced: a high-tension steel wire at the centre, inside a tubular copper conductor, provided strength. For the deep-sea section this meant the outer steel wires could be eliminated).

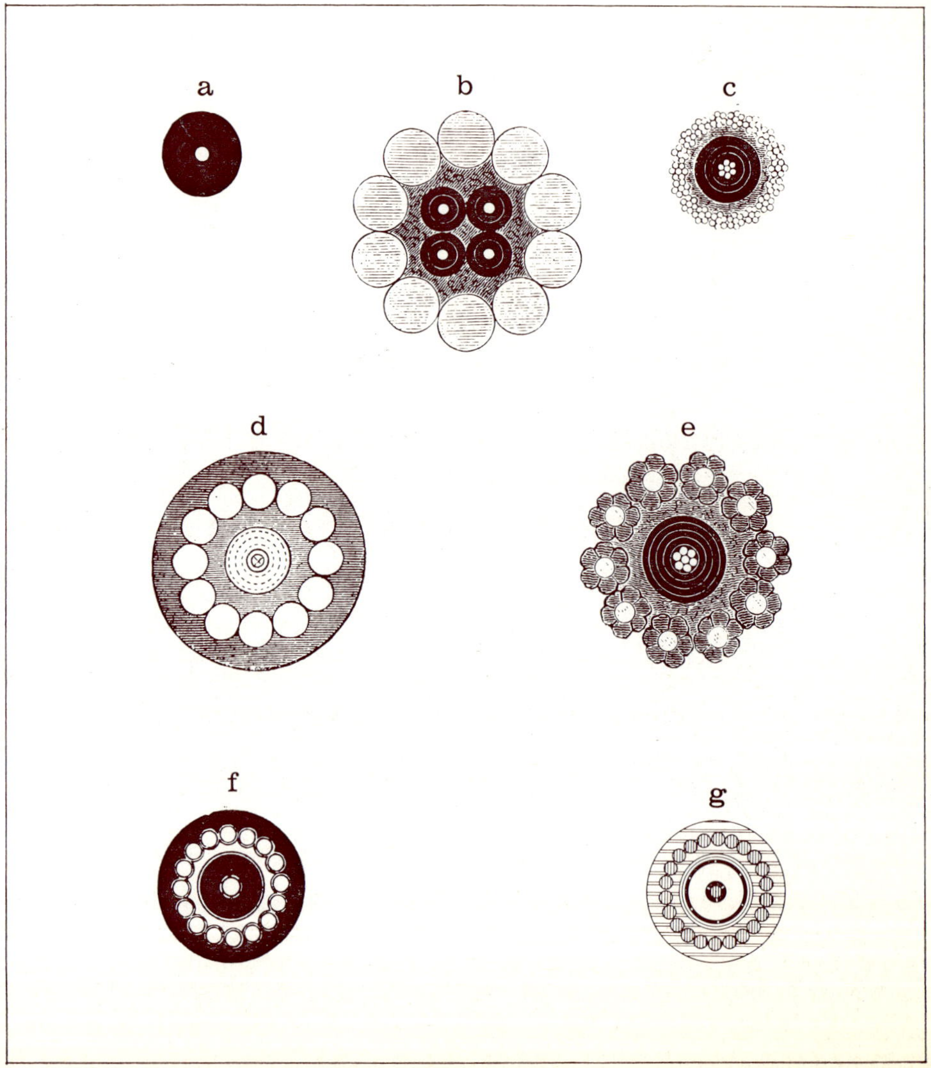

33 Cross sections of some submarine telegraph cables shown at actual size and described from the centre outwards.

3 The Mature Industry

With the completion of the Persian Gulf and Atlantic cables the industry passed out of its uncertain formative stages. The centre of activity, however, remained in Britain. There was a spurt of activity in the 1870s, financed in part by money that was freed when the Government nationalized the inland telegraph systems. Many of the new lines began to be consolidated under John Pender's guidance into the Eastern Telegraph Company, which would become the giant of the industry. Most of the cables continued to be manufactured on the banks of the Thames, principally by TC&M but also by Henley and by Siemens Brothers.

Basic cable design remained the same for almost a century: a stranded copper core covered with gutta percha insulation and then with iron armouring. But there were also some important changes, some of which will be noted here.

India rubber was occasionally used for insulation, especially where heat was a factor; and, as the quantities of material needed became great, gutta percha was combined with a similar material from South America called balata. In 1935, however, a synthetic plastic, polythene, was developed at ICI by the polymerisation of ethylene. In all essential respects it turned out to be as good as or better than gutta percha for cables. Polythene was widely used for cables during the war and since then has been used almost exclusively.

After 1879 brass tape was often added round the core, especially on cables laid in warm waters, for protection against boring insects.

In the 1880s Oliver Heaviside calculated that if inductance was 'loaded' on the cable the signal distortion due to capacitance would be reduced. Experiments with discrete loading coils were successfully performed on short lines after the turn of the century, but this was an impractical method for long lines. What was needed was a material with high magnetic permeability which could be wrapped continuously around the cable core. Two such materials were produced in the 1920s: an iron nickel alloy called *permalloy* by Western Electric in America, and a copper-iron-nickel alloy by TC&M which they called *mu-metal*.

34 Landing the shore end of the Pacific cable, 1902, with the Cable Ship Faraday in the background. The buoys will be cut loose when the cable is in position. The techniques involved in this operation have remained virtually unchanged for a hundred years

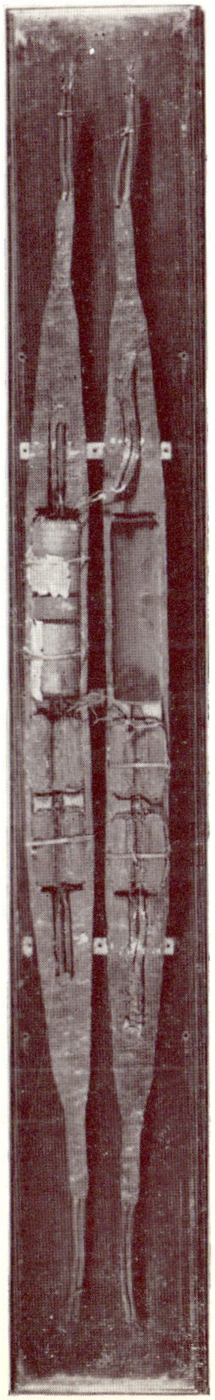

35 Coils used in the first discretely-loaded undersea cable–across the Channel–in 1910. An earlier cable using discrete (or 'lump') loading had been laid across Lake Constance between Germany and Switzerland in 1906

The first loaded cable was laid in the Atlantic in 1924, and others followed. Signal speeds were as much as ten times those of non-loaded cables.

The form of the conductor was also eventually changed. For the telegraph speeds being discussed the electrical resistance of the return path (through the ocean or through the cable armouring and brass tape) is not very significant. But it becomes significant at frequencies required for telephone transmission. This, added to the capacitance effect, placed severe restrictions on the possibilities of telephone transmission under water. The first telephone cable was laid across the Channel in 1891 as part of a London-Paris circuit. It had four cores, giving two telephone channels. Inductive loading was first used by C. E. Krarup of the Danish Telegraph Service on a line between Denmark and Sweden in 1902, and various other short telephone cables were designed and laid in the years following. Improvements were made by reducing the capacitance, increasing the inductive loading, or lowering the resistance. In 1921 three coaxial cables, each consisting of a central copper wire and an outer copper tube separated by gutta percha and loaded with soft iron wire, were successfully laid between Florida and Cuba.

Plans were made in 1928 to lay a loaded transatlantic telephone cable, using coaxial conductors and paragutta insulation (paragutta was a mixture of de-resinated gutta percha, or balata, deproteinized rubber, and petroleum wax). But the advent of short-wave radio, combined with the economic depression of the thirties, caused the project to be dropped.

Transmitting and receiving equipment evolved slowly over the years. Special keys were designed for curb transmission, described by William Thomson and Fleeming Jenkin in an 1861 patent. Curbing employs a reverse pulse transmitted immediately after the main pulse to help sharpen the received signal. In the 1880s automatic machines fed by a punched tape were constructed, and after 1900 the punched tape was being produced by keyboard perforators.

36 An early Thomson siphon recorder

37 Heurtley magnifier

Problems at the receiving end were more complex. Thomson designed a 'siphon recorder' which was less sensitive than the mirror galvanometer but had the advantage that the signal was permanently recorded on paper. Variations of this instrument were widely used from the 1870s.

But someone still had to read the tape and retransmit the message, a time consuming and error-prone exercise. Therefore much effort was spent in trying to design instruments that would relay the message automatically. To do this the signal first had to be amplified, or 'magnified'. One popular magnifier, widely used on Western Union lines, was developed by E. S. Heurtley in 1908. The moving coil, similar to that of the siphon recorder, was attached to two wires. The wires were heated by the electric current flowing through them and cooled by two small blowers. When the coil turned, one wire moved closer to its blower, the other moved away. Since resistance depends on temperature, the resistance of the first wire decreased and that of the other increased. This upset the balance of an electrical circuit of which the two wires were a part, and produced an amplified signal. The output of the Heurtley magnifier was large enough to operate a relay which produced a signal strong enough to send on to the next station. One form of relay, designed by S. G. Brown, also operated on the moving coil principle; the coil moved a needle which rested on a rotating drum (rotating in order to reduce the friction). There were two conducting bands on the drum, separated by a neutral area, and as the needle moved back and forth under the influence of the coil it would make contact first with one band, then with the other. The Heurtley magnifier and Brown drum relay came into widespread use after the first world war.

Methods of sending signals on landline telegraph wires in opposite directions at the same time, called duplex telegraphy, were developed in the 1850s. Application to submarine systems was more difficult because of the capacitance of the cables, but even this was done successfully in the 1870s, first by J. B. Stearns, then by H. A. Taylor and Alexander Muirhead. The Taylor-Muirhead method, which was

38 Brown drum relay

39 Western Union distributor

40 Capacity magnifier

widely used, required construction of an 'artificial line', using resistors and capacitors to duplicate the characteristics of the real line. The cable and the artificial line were then treated as two parts of a balanced circuit (Wheatstone bridge). At the transmitter end the signal would go out through both lines in such a way that the effects would cancel out in the local receiver, and all the receiver would 'see' would be the incoming signal.

As higher and higher transmission speeds became possible, Western Union practice on its Atlantic cable – where a great deal of traffic occurred between two points – was to run several channels at relatively low speeds of about 50 words per minute rather than using one high speed channel; the speed was chosen to correspond with what an individual operator could type and what the Heurtley magnifier and printers could receive. The time-sharing channel apparatus consisted of a distributor with contact brushes rotating over a segmented face plate; each segment corresponded to a channel. Synchronization was established with the help of electrically driven tuning forks at the two ends of the line.

Equipment and practice for the Eastern system developed somewhat differently. The capacitance magnifier was widely used on long lines. A moving coil altered the distance between plates in a differential capacitor, increasing the capacitance on one side and decreasing it on the other. This modified the input signals differently for two valve circuits. The difference was amplified by the valves and the output was fed into a relay, usually a 'fork relay'. This was also a moving-coil instrument with the coil attached to two wires making contact with stationary pins. In contrast to Western Union's situation on the North Atlantic, the Eastern lines did not have heavy traffic demands between any two points. There was therefore no need for many-channelled systems. Two channels, however, were installed on a number of lines by means of cam-operated contacts. This type of operation was especially suitable for some of the long chains in the Eastern system (a message from London to the Far East or Australia, for instance, could pass through a dozen relay points).

41 The deck of the *Great Eastern*. The dynamometer (left of centre, facing the men) measures the strain in the cable

Electronic amplifiers were developed in the 1920s and replaced some of the electro-mechanical magnifiers. But there were strong feelings in some quarters for the reliability of the old instruments, which continued to be used in many stations.

Finally, techniques of cable laying have not changed greatly since the 1860s. Better soundings of the ocean bottom have given greater confidence in the laying operation, and experience on deck has led to modifications of the equipment used there. William Siemens introduced the use of a taut steel wire to be paid out along with the cable but separate from it; the difference between the lengths of the two gave the amount of slack in the cable. It was then possible to judge whether there was enough cable to allow for the hills and valleys on the ocean bottom. The system was first installed on the *Faraday* in 1874. In more recent times an important development has been the introduction of ploughing. Damage to cables from ships' anchors and fishing gear has always been a severe problem, as has chafing from tidal motion – which is why sections in shallow waters have been heavily armoured. It was long recognized that an even better solution would be to bury the cable under the seabed. A suitable plough was developed by Western Union, and a few shore ends were successfully buried as early as 1938.

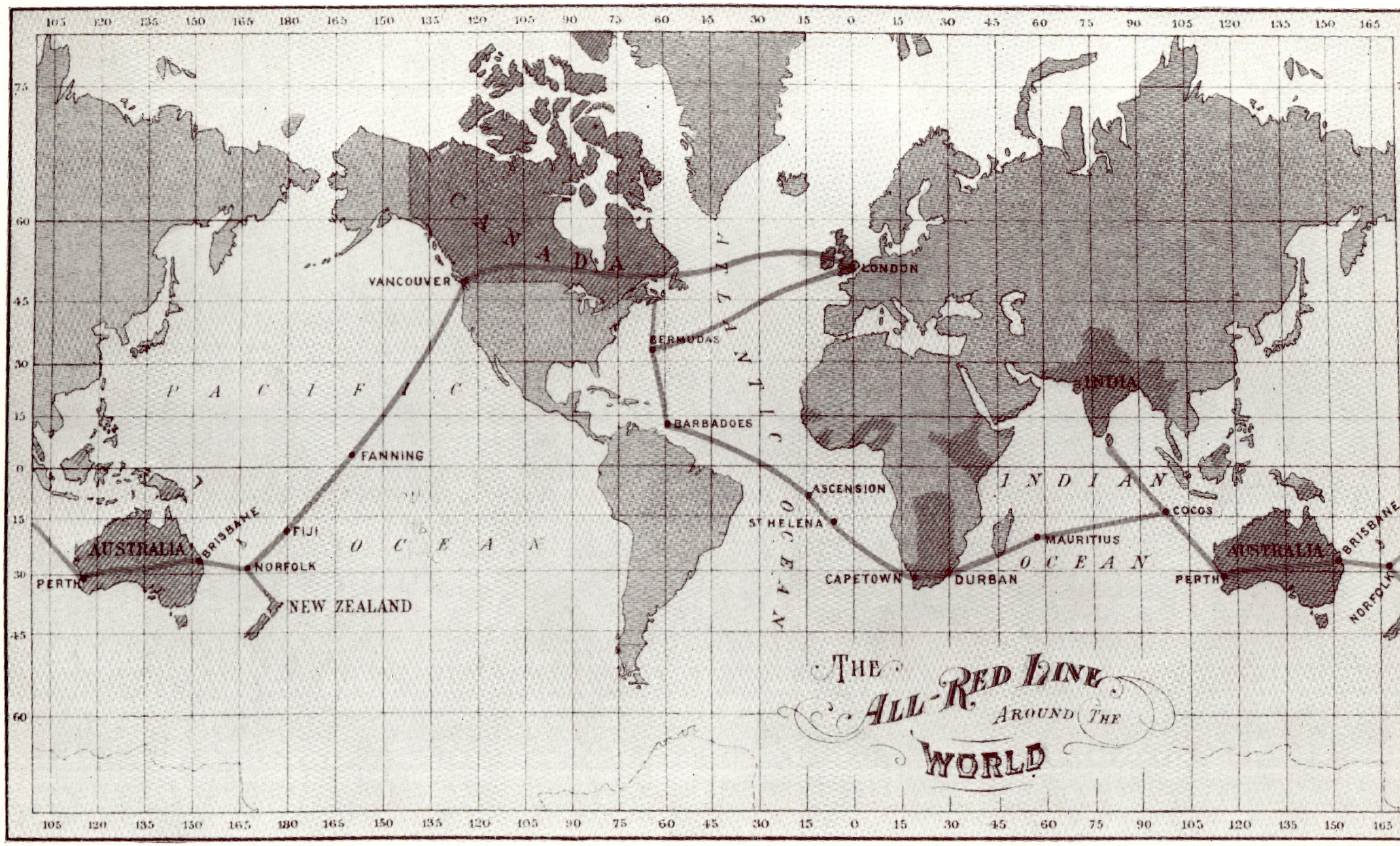
The All-Red Line Around the World
CANADA
VANCOUVER
LONDON
BERMUDAS
BARBADOES
FANNING
FIJI
NORFOLK
BRISBANE
AUSTRALIA
PERTH
NEW ZEALAND
ASCENSION
ST HELENA
CAPETOWN
DURBAN
MAURITIUS
COCOS
INDIA
PACIFIC OCEAN
ATLANTIC OCEAN
INDIAN OCEAN

42

EFFECT OF THE SUBMARINE TELEGRAPH; OR, PEACE AND GOOD-WILL BETWEEN ENGLAND AND FRANCE.

43 Cartoon from Punch, 1851

4 Consequences of the cables

In addition to being influenced by economic consideration, the early cable promoters were motivated by nationalistic feelings. As already mentioned, there was a strong patriotic element in the promotion of British cables. From this developed the concept that all sections of the basic system should be kept in British hands. Hence in 1902 the Pacific cable from Vancouver to Australia and New Zealand went by way of Fanning Island instead of making a more practical stop at Hawaii; this cable was the final link in an 'all red' (all British) line around the world.

Other governments, especially those with overseas possessions, responded in like manner. Many articles were written towards the close of the nineteenth century and in the early years of the twentieth declaiming upon the strategic importance of national cables in time of war and upon their general political importance in time of peace. These motivations were important factors in the laying of several cables, especially those promoted by German, French and American interests.

At the same time, of course, many cables were links between independent nations, the statement often being made that improved communications improved the chances for peace. The opinion was expressed that one good cable was worth a dozen peace conferences.

Some of these expectations have clearly been fufilled. The telegraph cables had a tremendous economic impact on the world in allowing trade to be conducted more efficiently. But they also accelerated the rate at which an economic disturbance in one part of the globe was felt in another, helping to make economic fluctuations sharper and more widespread.

Politically the cause of peace has no doubt often been served by cables. Diplomats have been able to consult immediately with their governments in moments of crisis, and misunderstandings have been avoided (although it had been said more than once that a little delay, and a little misunderstanding, might sometimes be a good thing; that some events which seem important at the moment have a way of seeming less so if a few days or weeks are allowed to pass). It should

452 THE ILLUSTRATED LONDON NEWS [MAY 12, 1855.

44 Landing the Crimean cable, 1855

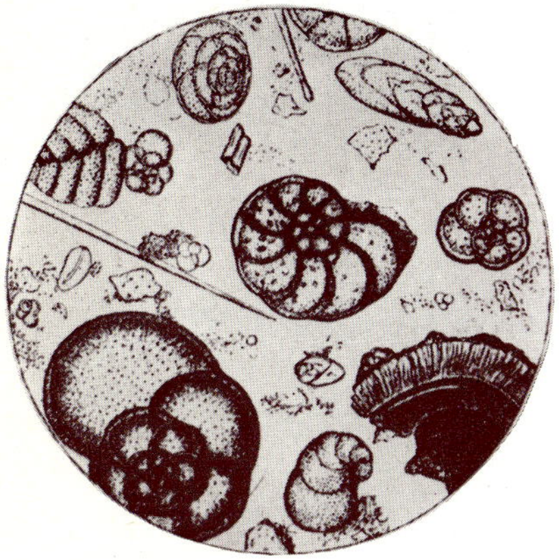

45 Sample from ocean bottom, 1856

also be pointed out that an ambassador who was able to communicate instantaneously with the central government no longer had such complete authority to act on his own; this had a marked effect on the conduct of foreign policy, the development of the bureaucracy at home, and on the type of person who sought ambassadorial posts.

The news services have from the beginning been prime users of submarine cables, as they were of land lines. Public awareness of and reaction to world events changed markedly as a result; so too has the character of the news services themselves.

The extent to which cables have promoted peace is difficult to assess. But they certainly have been useful in time of war. A special cable was laid across the Black Sea to the Crimea in 1855 and was in continual use during that ill-fated venture. In both world wars cables were cut by the opposing sides in attempts to disrupt communications. Having better control of the seas, the Allies were not only more successful in cutting operations, they even managed to divert some of the German cables (in both wars) and put them to their own use.

There have also been important technical and scientific side effects from cable activities. Soundings made for cables gave the first evidence of the existence of life at great depths and helped to stimulate a surge of interest in oceanography in the late nineteenth century. The need for more precise electrical definitions in submarine telegraphy caused Latimer Clark and Charles Bright to present a paper to the British Association in 1861 which led to the formation of the Committee on Electrical Standards. The Committee sponsored some exacting experimental work and produced definitions of the basic electrical units. The cables were of immediate value in tying observatories together, allowing them to determine their longitudes more precisely and to coordinate observations.

46 Repeater being launched from cable ship *Monarch*, 1862

5 The sudden end and a new beginning

In the 1920s radio effectively challenged the cables over many routes. Combined with the economic problems of the following decade this new development helped to slow down the expansion of cable networks. However, cables were more reliable than radio; they also could be more confidential. Much information therefore continued to flow through them.

After the second world war, however, work accelerated on a development that would spell the end to key parts of the old cable technology. Repeaters were designed which were small enough to be laid with the cable and reliable enough to be left submerged under water without failing over long periods of time. Several successful applications were made on short lines. Then, in 1956, the first transatlantic repeatered cable was laid, between Clarenville, Newfoundland, and Oban, Scotland. There were actually two cables on this section of the route, each with 51 single-directional repeaters, and each capable of carrying thirty-six 4kHz telephone channels (later changed to forty-eight 3kHz channels, with additional telephone links possible by a technique that uses the pauses that occur in ordinary conversation). A single cable with twelve two-way repeaters completed the system to the North American mainland. This new cable system could carry more information than all the old ones put together (the best of the older transatlantic cables were operated at band-widths of less than 100Hz); and there was good reason to believe this was only the beginning, as indeed it was. The communications satellites which began to appear in the following decade had even greater band-width capabilities, to the extent of being able to transmit television signals.

The result has been that less than two decades after the laying of the first transatlantic cable the old long distance telegraph cables have been abandoned, the specialized equipment dispersed and the techniques discarded. Shorter lines remain only where the demands for communications capabilities are minimal and where telephone cables or microwave facilities are not economical.

The new communications era, like the old one, was inaugurated with

47 Cable network in 1973

forecasts of great expectations: the world would become a more integrated place as people, businesses and governments were better able to communicate. This would lead to more mutual understanding, and a friendlier, more peaceful world. It will still be some time before proper historical judgements can be made on these aspirations.

Further Reading

Charles Bright, *Submarine Telegraphs* (London, 1898)
Samuel Carter, *Cyrus Field, Man of Two Worlds* (New York, 1968)
Bern Dibner, *The Atlantic Cable* (Norwalk, 1959)
Gerald Garratt, *One Hundred Years of Submarine Cables* (London, 1950)
Kenneth Haigh, *Cable Ships and Submarine Cables* (London, 1968)

Sources of Illustrations

British Museum 8
C. Bright, *Submarine Telegraphs* 18, 19, 21, 26
Cable & Wireless 13, 16, 40, 41, 46
Harpers Weekly 7
Illustrated London News 1, 20, 22, 23, 32, 44, 45
L'Illustration 24
Institution of Electrical Engineers 11, 14, 18a, 19, 21, 23, 26, 29, 30
A. G. King, *Kelvin the Man* 15
Post Office Museum 34
Punch 2, 43
Smithsonian Institution 7
E. Zetzsche, *Handbuch der Elektrischer Telegraphie* 26a, b, c, e

The other illustrations are Science Museum photographs.